Diabetes

The Complete Guide on Type 1 & Type 2 Diabetes, Signs, Causes and Treatments. A Long-Lasting Solution to Diabetes Mellitus with the Help of Medication, Self-Monitoring and Healthy Diets.

Doctor Jane A. McCall

Table of Contents

Diabetes 1

INTRODUCTION 12

CHAPTER 1 15

  WHAT IS DIABETES? 15

CHAPTER 2 18

  WHAT IS PRE-DIABETES 18

    Symptoms of Pre-Diabetes 19

    Causes and Risk Factors of pre-diabetes 21

    Diagnosis of Pre-Diabetes 25

    How to Treat Pre-Diabetes 29

CHAPTER 3 34

  WHAT IS TYPE 1 DIABETES? 34

    Causes of Type 1 Diabetes 36

    Common Symptoms and Signs Of Type 1 Diabetes. 38

    Life with Type 1 Diabetes 40

CHAPTER 4 42

  WHAT IS TYPE 2 DIABETES? 42

    Causes of Type 2 Diabetes 44

Common Diabetes Symptoms Associated With Type 2 Diabetes. 46

CHAPTER 5 49

WHAT IS GESTATIONAL DIABETES? 49

Maturity Onset Diabetes of The Young 51

CHAPTER 6 53

COMPLICATIONS OF BADLY CONTROLLED DIABETES 53

Symptoms Caused by Diabetes Complications 57

Symptoms Related To Nerve Damage (Neuropathy) 58

Skin-Related Diabetes Symptoms: 61

Eye-Related Diabetes Symptoms 63

CHAPTER 7 66

CONTROLLING DIABETES-TREATMENT IS EFFECTIVE AND IMPORTANT. 66

Natural Ways to Help Control Diabetes Symptoms 68

Physical state of being inactive and weight 71

CHAPTER 8 76

WHAT IS HYPOGLYCEMIA? 76

Hypoglycemia Symptoms and Signs 80

    Hypoglycemia Causes and Risk Factors 84

    Conventional Treatment for Hypoglycemia...........89

    Facts about Hypoglycemia 96

    Precautions Regarding Hypoglycemia 102

    Final Thoughts on Hypoglycemia 103

CHAPTER 9...106

  WHAT IS HYPERGLYCEMIA? 106

    Symptoms of Hyperglycemia................................107

    Causes of Hyperglycemia......................................108

    Treatment and Prevention Of................................109

    Hyperglycemia 109

    Hyperglycemia Can Lead To 111

    Diabetic Ketoacidosis..111

CHAPTER 10...114

  WHAT IS INSULIN?..114

    Taking Insulin 116

    Different Preparations of Insulin..........................117

    Practical Advice for Injecting Insulin 121

    Insulin Side Effects 123

    Inhaled Insulin..126

CHAPTER 11 130

INSULIN PUMP FOR CONTROLLING DIABETES. 130

   Insulin Pumps for Type 1 131

   Diabetes. 131

   INSULIN PUMPS FOR TYPE 2 134

   DIABETES 134

   Future Options for Delivering 135

   Insulin 135

   Artificial Pancreas 136

CHAPTER 12 138

   WHAT IS ORAL INSULIN? 138

   Treatments for Diabetes 140

   How do oral drugs lower glucose levels? 141

   What oral drugs are available for type 2 diabetes? 143

CHAPTER 13 149

   5-STEP PLAN TO REVERSE DIABETES 149

   Step 1: Remove These Foods to Reverse Diabetes Naturally. 149

   Step 2: Incorporate These Foods to Treat Diabetes 153

   Step 3: Take These Supplements for Diabetes 157

Step 4: Follow This Eating Plan to Reverse Diabetes 161

Step 5: Exercise to Balance Blood Sugar 165

Final Thoughts on How to Reverse Diabetes 167

CHAPTER 14 170

SELF-MONITORING OF BLOOD GLUCOSE 170

What Is Blood Glucose Self-Monitoring? 171

Who Should Self-Monitor Blood Glucose? 173

Target Blood Glucose Levels 175

How Is A Blood Glucose Monitor Used? 177

When Should Glucose Self-Monitoring Tests Be Done? 180

Real-Time Continuous Glucose Monitoring 182

CHAPTER 15 184

MANAGING DIABETES WITH DIET & FOOD PLANNING. 184

What DIET IS BEST FOR DIABETES? 185

Professional Help with Lifestyle Changes For Diabetes 190

Obesity, Diabetes and Diet 191

CHAPTER 16 193

MANAGING DIABETES WITH PHYSICAL

ACTIVITY AND EXERCISE 193

    Exercise and Diabetes 194

    Types of Exercise For People With Diabetes.......197

    Monitoring Blood Glucose Levels When Exercising 203

    Trying Out Blood Glucose Before, At Some Stage In, And After Exercising 204

    Hypoglycemia and Exercise..................................207

    When to See a Doctor...209

    Other Considerations..211

Recommendations 213

About the Author 217

Acknowledgments 218

INTRODUCTION

Have you been diagnosed with diabetes or have a pre-diabetic condition, or think you might be developing the disease, or just want to know a bit more about the condition and how to avoid or control it, here are some basic statistics about the disease that you might find interesting. Diabetes, also known as diabetes mellitus, is a chronic metabolic condition in which the sufferer has high levels of blood glucose (sugar). This is due to the pancreas either not producing enough insulin, or because the body does not respond to the insulin that is produced. It is an increasingly common condition, and statistics from the World Health Organization reports that there are over 346 million diabetes sufferers worldwide, and that 90% suffer from Type 2 diabetes.

This book has help millions of people in the world to discover Problems and complications associated with diabetes. Side effects of the disease can include heart disease, stroke, high blood pressure, eye problems and blindness, kidney disease, complications in pregnancy, and mental health problems such as depression. The aim of this book is to provide some basic and useful information about diabetes, particularly type 1 and type 2 diabetes, likewise you will discover what diabetes is, what causes diabetes, how it might be prevented, how it can be controlled, how to avoid the problems and complications associated with the disease, and how to cope with the disease so that you can live a long and fulfilling life.

CHAPTER 1

WHAT IS DIABETES?

Diabetes, frequently known as diabetes mellitus, is a group of metabolic diseases in which the victim has excessive blood glucose (blood sugar), either due to the fact insulin production is inadequate, or due to the fact that the cells of the body's fail to respond well to insulin, or both. Patients with excessive (high) blood sugar will frequently experience polyuria (common urination), they may end up more and more thirsty (polydipsia) and increased hunger (polyphagia). If left untreated, diabetes can cause many complications. Acute complications can include diabetic ketoacidosis, hyperosmolar hyperglycemic state, or demise. Serious long term complications encompass cardiovascular sickness, stroke, chronic kidney disorder, foot ulcers, and harm to the eyes. Diabetes is due to either the pancreas not producing enough insulin or the cells of the body not responding properly to the insulin produced. The fundamental types of diabetes mellitus:

- Type 1 DM (diabetes mellitus): results from the pancreas's failure to produce enough insulin. This form was formally referred to as "insulin-dependent diabetes mellitus" (IDDM) or "juvenile diabetes".

- Type 2 DM (diabetes mellitus): starts off with insulin resistance, a situation in which cells fail to reply to insulin properly. as the disorder progresses a loss of insulin may additionally expand. This shape was formerly called "non-insulin-dependent diabetes mellitus" (NIDDM) or "person-onset diabetes".

- Gestational diabetes: is the third fundamental form and this do occurs when pregnant women without a preceding history of diabetes have high blood sugar levels due to one factor or the other.

CHAPTER 2

WHAT IS PRE-DIABETES

Pre-diabetes is the precursor stage before diabetes mellitus wherein not all the signs and symptoms required to diagnose diabetes are present, however blood sugar is abnormally high. This stage is frequently referred to as the "grey area." it is not always a disease likewise pre-diabetes is a "pre-analysis" of diabetes—you could consider it as a warning sign. It's while your blood glucose level (blood sugar level) is higher than what it ought to be, but it's not high enough to be taken into consideration as diabetes.

Pre-diabetes is a sign that you may develop Type 2 diabetes in case you don't make a few lifestyle modifications.

It's quite possible to prevent pre-diabetes from growing into type 2 diabetes. Eating healthy food, losing weight and staying at a healthy weight, and being bodily active can help you bring your blood glucose level reduced back into the normal stage.

Symptoms of Pre-Diabetes

Diabetes develops very progressively, so while you're in the pre-diabetes stages-when your blood glucose level is higher than it ought to be-you could no longer have any symptoms in any respect. You can, however, notice that:

- You're hungrier than normal
- You're dropping weight, in spite of ingesting more
- You're thirstier than normal
- You have to visit the rest room more frequently
- You're more worn-out than regular

All of these are normal signs and symptoms related to diabetes, so if you're in the early degrees of diabetes, you may note them.

Causes and Risk Factors of pre-diabetes

Pre-diabetes develops while your body starts to have hassle using the hormone insulin. Insulin is important to transport glucose-what your body uses for energy-into the cells through the bloodstream. In pre-diabetes, it shows that your body doesn't make enough insulin or it doesn't use it properly (that's referred to as insulin resistance).

If you don't have sufficient insulin or if you're insulin resistant, you can build up too much glucose for your blood, leading to a higher than ordinary blood glucose level and possibly pre-diabetes.

Researchers aren't sure what precisely causes the insulin process to go awry in some people. There are several risk elements, though, that make it much more likely that you'll develop pre-diabetes. Those are the identical danger factors related to the development of type 2 diabetes:

1. **Weight:** in case you're overweight (have a body mass index-a BMI-higher than 25), you're at an excessive risk for developing pre-diabetes. Most especially if you carry quite a few extra weight in your abdomen, you may develop pre-diabetes. The more fat cells can cause your body to end up more insulin resistant.

2. **Lack of physical activity:** this often goes hand-in-hand with being overweight. If you aren't physically active, you're much more likely to develop pre-diabetes.

3. **Family history:** pre-diabetes has a hereditary component. If someone in your own family has (or had) it, you're more likely to develop it.

4. **Race/ethnicity:** certain ethnic groups are much more likely to increase pre-diabetes, such as Africans-Americans, Hispanic Americans, Native Americans, and Asian Americans.

5. **Age:** the older you are, the more at risk you're for developing pre-diabetes. At age forty-five, your chance

starts to rise, and after age sixty-five, your chance increases exponentially.

6. **Gestational diabetes:** if you develop diabetes as at the time you were pregnant, that will increase your risk for developing pre-diabetes afterward.

7. **Other health problems:** excessive blood pressure (hypertension) and high cholesterol (the "bad" LDL cholesterol) increase your hazard of having type 2 diabetes.

Polycystic ovary syndrome (PCOS) additionally raises the chance for pre-diabetes as it is associated with insulin resistance. In PCOS, many cysts form in your ovaries, and one possible reason is insulin resistance. When you have PCOS, which means you'll be insulin resistant and consequently at risk for developing pre- diabetes.

Diagnosis of Pre-Diabetes

Your doctor may need to check your blood glucose level in case you're overweight (have a body mass index-BMI- of over 25) and when you have one or more of the risk factors indexed above.

Even if you are not overweight and don't have any of the risk factors, your health practitioner might also need to begin testing out your blood glucose level every three years beginning while you're forty-five. That's a smart factor to do because the risk of developing pre-diabetes (and consequently type 2 diabetes) increases with age. Due to the fact that there are so many possible complication of diabetes (e.g., heart problems and nerve issues), it's an awesome idea to be vigilant about detecting blood glucose abnormalities early.

To diagnose you with pre-diabetes, the doctor can run one of two tests-or he/she may decide to do both. The checks are:

Fasting plasma glucose test (FPG): you won't eat anything for 8 hours leading up to a FPG take a look at. That's why a FGP test is often done in the morning. The doctors exams your blood glucose level (blood sugar level) after drawing a small blood sample.

In case your blood glucose level is among 100 and 125mg/dl, you have pre-diabetes. You may hear the doctor use the word "impaired fasting glucose" or IFG, that is another term for pre-diabetes whilst it is identified with the fasting plasma glucose test.

In case your blood glucose level is above 126mg/dl with the FGP test, you may have diabetes.

Oral glucose tolerance test (OGTT): that is some other test used to diagnose pre-diabetes. The medical doctor will give you instructions on the way to prepare for the test, but you won't be able to eat anything for 8 hours earlier than the test; you'll be fasting. In that way, the oral glucose tolerance test, abbreviated OGTT, is much like the fasting plasma glucose test.

On the day of the test, the health practitioner will test your blood glucose level at the beginning of the appointment; that's known as your fasting blood glucose level. Then, you'll drink 75g of a completely sugary mixture. Two hours later, your blood glucose degree will be measured.

In case your blood glucose degree is between 140 and 199mg/dl two hours after ingesting the sugary combination, you have got pre-diabetes. You may hear the doctor use the word "impaired glucose tolerance" or

IGT, which is another term for pre-diabetes while it's diagnosed with the OGTT.

If your blood glucose level is above 200mg/dl with the oral glucose tolerance test, you may have diabetes.

How to Treat Pre-Diabetes

The American diabetes association says that several lifestyle changes are powerful in preventing Type 2 diabetes once you've been diagnosed with pre-diabetes. Your doctor will walk you through what you need to change, however typical recommendations are:

1. **Eat properly:** a registered dietitian (RD) or licensed diabetes educator (CDE) will help you create a meal plan that's full of good-for-you and good-for-your-blood-glucose-level meals. The intention of the meal plan is to control your blood glucose level and keep it in the healthy, normal range. Your meal plan will be made just for you, taking into account your overall health, physical activity, and what you like to eat.

2. **Exercise:** when you exercise, your body uses more glucose, so exercising can reduce your blood glucose level. In addition while you exercise, your body doesn't need much insulin to transport the glucose; your body becomes much less insulin resistant. Since your body

isn't using insulin properly if you have pre-diabetes, a decrease insulin resistance is a very good thing.

And of cause, there are all of the conventional benefits of exercise: it can help you lose weight, keep your coronary heart healthy, make you sleep better, and even enhance your mood.

The American diabetes association recommends at the least a 150minutes of moderate activity per week—that's 30minutes five days a week. You could get that through sports consisting of walking, bike using, or swimming.

3. **Lose weight:** If you're overweight, you should get started on a weight loss as soon as you're diagnosed with

pre-diabetes. Losing simply 5-10% of your weight can significantly reduce your hazard of developing Type 2 diabetes. The aggregate of consuming nicely and exercise more is a splendid way to lose weight-after which you maintain your new, healthful weight.

4. **Metformin:** for peoples that are at a very high risk of developing Type 2 diabetes after being diagnosed with pre-diabetes, the doctor may suggest a medication. The American diabetes association says that metformin should be the handiest medication used to prevent type 2. It really works by retaining the liver from making greater glucose when you don't want it, thereby maintaining your blood glucose level in a better range.

Your health practitioner will keep a near watch on your blood glucose level, monitoring them to make sure that your pre-diabetes doesn't end up Type 2 diabetes. If needed, he or she may suggest adjustment changes (e.g., different diet or more exercise) to better control your blood glucose levels.

CHAPTER 3

WHAT IS TYPE 1 DIABETES?

Type 1 diabetes is an autoimmune ailment wherein the immune gadget destroys cells inside the pancreas. The body does no longer produce insulin which will help to fight against any form of infection or disease. Normally, the disease first appears in adolescence or early adulthood. Type 1 diabetes used to be called juvenile-onset diabetes or insulin-based diabetes mellitus (IDDM), however the disease could have an onset at any age. Type 1 diabetes makes up round 5% of all instances of diabetes.

In Type 1 diabetes, the pancreas is unable to produce any insulin, the hormone that controls blood sugar levels. Insulin production turns into inadequate for the control of blood glucose levels due to the sluggish destruction of beta cells inside the pancreas. This destruction progresses without been aware over time until the mass of those cells decreases. To the extent that the quantity of insulin produced is insufficient.

Type 1 diabetes generally appears in childhood or youth, however its onset is likewise possible in maturity.

While it develops later in lifestyles, type 1 diabetes can be mistaken for type 2 diabetes. If correctly diagnose, it's miles referred to as latent autoimmune diabetes of maturity.

Causes of Type 1 Diabetes

The gradual destruction of beta cells in the pancreas that sooner or later consequences within the onset of type 1 diabetes is the result of autoimmune destruction. The immune machine turning against the body's own cells is probably brought on by means of an environmental component uncovered to human beings who've a genetic susceptibility.

Despite the fact that the mechanisms of Type 1 diabetes

etiology are uncertain, they're thought to contain the interaction of more than one factors which are as follows:

- Susceptibility genes - some of which might be carried by over 90% of patients with type 1 diabetes. A few populations - scandinavians and sardinians, for example - are much more likely to have susceptibility genes.

- Autoantigens - proteins thought to be launched or exposed at some point of ordinary pancreas beta cell turnover or Damage along with that caused by infection. The autoantigens set off an immune response resulting in beta cell Destruction.

- Viruses - coxsackievirus, rubella virus,

cytomegalovirus, epstein-barr virus and retroviruses are amongst the ones which have been connected to type 1 diabetes.

- Eating regimen(Diet)- toddler exposure to dairy products, high nitrates in consuming water and low vitamin D intake have additionally been connected to the development of type 1 diabetes.

Common Symptoms and Signs Of Type 1 Diabetes.

- Regularly feeling thirsty and having a dry mouth
- Changes on your urge for food, commonly feeling very hungry, occasionally even if you've currently

eaten (this could also occur with weak point and problem Concentrating)

- Fatigue, feeling usually tired no matter slumbering and mood swings
- Blurred, worsening vision
- Gradual recuperation of skin wounds, common infections, dryness, cuts and bruises
- Unexplained weight changes, specifically dropping weight despite consuming the identical quantity (this occurs due to the body using alternative fuels stored in muscle and fat whilst releasing glucose in the urine)
- Heavy respiratory

- Potentially a loss of recognition

- Nerve damage that causes tingling sensations or ache and numbness in the limbs, feet and arms (more common among people with Type 2 Diabetes)

Life with Type 1 Diabetes

Type 1 diabetes continually requires insulin treatment and an insulin pump or day by day injections could be a lifelong requirement to keep blood sugar levels under control. The situation was known as insulin structured diabetes.

After the prognosis of type 1 diabetes, health care

providers should help patients discover ways to self-monitor via finger stick testing, the signs of hypoglycemia, hyperglycemia and different diabetic complication. Most patients will also be taught how to regulate their insulin doses.

As with other forms of diabetes, vitamins and physical hobby and workout are vital factors of the way of life control of the sickness.

CHAPTER 4

WHAT IS TYPE 2 DIABETES?

Type 2 diabetes is the most popular form of diabetes, accounting for over 90% of all diabetes cases.

The number of adults identified with diabetes within the US has risen notably inside the past 30 years, almost quadrupling from 5.5 million cases in 2000 to more than 21.3 million in 2017.

Type 2 diabetes was once called adult-onset diabetes and Noninsulin-structured diabetes mellitus (NIDDM), however the ailment may have an onset at any age, increasingly more along with early life.

Type 2 diabetes mellitus most commonly develops in adulthood and is more likely to occur in folks that are overweight and physically inactive.

Not like Type 1 diabetes which presently cannot be prevented, a number of the risk factors for type 2 diabetes may be modified. For lots of people, consequently, it is viable to save you the circumstance. Signs and symptoms that signal the need for diabetes testing:

- Frequent urination
- Weight reduction
- Loss of energy
- Excessive thirst.
- Slow healing of cuts

- Numbness or tingling in hands and feet

- Itchy skin

Causes of Type 2 Diabetes

Insulin resistance is normally the precursor to type 2 diabetes - a situation in which more insulin than normal is needed for glucose to enter cells. Insulin resistance inside the liver results in more glucose production at the same time as resistance in peripheral tissues means that glucose uptake is impaired. The impairment stimulates the pancreas to make extra insulin however sooner or later the pancreas is unable to make sufficient to save your blood sugar levels from growing too excessive.

Genetics performs a part in Type 2 diabetes - relatives of people with the disease are at a higher risk, and the Prevalence of the condition is much higher in particular amongst local Americans, Hispanic and Asian human beings.

Obesity and weight gain are vital elements that lead to insulin resistance and Type 2 diabetes, with genetics, diet, exercise and way of life all playing an element. Body fat has hormonal effects on the impact of insulin and glucose metabolism.

Once type 2 diabetes has been identified, health care

provider can help patients with a program of education and monitoring, including the way to spot the signs of hypoglycemia, hyperglycemia and other diabetic intricates.

As with other kinds of diabetes, nutrients, and bodily pastime and exercise are critical elements of the life-style Management of the situation.

Common Diabetes Symptoms Associated With Type 2 Diabetes.

Many people develop type 2 diabetes symptoms in midlife or in older age and gradually expand signs in stages, especially if the condition is going untreated and worsens. Type 2 diabetes signs and symptoms can consist of:

- Chronically dry and itchy skin
- Patches of dark, velvety skin inside the folds and creases of the body (normally in the armpits and neck). This is known as acanthosis nigricans.
- Common infections (urinary, vaginal, yeast and of the groin)
- Weight benefit, even without a change within the diet
- Ache, swelling, numbness or tingling of the hands and toes

- Sexual disorder, consisting of loss of libido, reproductive issues, vaginal Dryness and erectile dysfunction.

CHAPTER 5

WHAT IS GESTATIONAL DIABETES?

Gestational diabetes mellitus (GDM) resembles Type 2 DM in several respects, regarding a mixture of enormously inadequate insulin secretion and responsiveness. It takes place in about 2–10% of all pregnancies and might improve or disappear after delivery.

But, after pregnancy about 5–10% of ladies with gestational diabetes are found to have diabetes mellitus, most normally Type 2.

Gestational diabetes is fully treatable, but calls for careful medical supervision throughout the pregnancy. Management may additionally encompass dietary changes, blood glucose tracking, and in a few instances, insulin can be required.

Though it may be temporary, untreated gestational diabetes can harm the fitness of the fetus or mother. Dangers to the baby include macrosomia (high birth weight), congenital coronary heart and central nervous system abnormalities, and skeletal muscle malformations.

Expanded tiers of insulin in a fetus's blood may additionally inhibit fetal surfactant production and cause respiratory distress syndrome. A high blood bilirubin stage may additionally result from red blood cell destruction. In intense instances, perinatal loss of life may occur, most commonly because of negative placental perfusion because of vascular impairment. labor induction may be indicated with reduced placental function. A caesarean section may be done if there is marked fetal distress or an increased hazard of damage related to macrosomia, along with shoulder dystocia.

Maturity Onset Diabetes of The Young

Maturity onset diabetes of the young (MODY) is an autosomal dominant inherited form of diabetes, due to several one of numerous single-gene mutations inflicting defects in insulin production. It's far drastically less not unusual than the 3 main types (Type 1, Type 2, and Gestational). The name of this sickness refers to early hypotheses as to its nature. Being due to a defective gene, this ailment varies in age at presentation and in severity in keeping with the precise gene disorder. Human beings with MODY regularly can manage it without the use of insulin.

CHAPTER 6

COMPLICATIONS OF BADLY CONTROLLED DIABETES

Below is a list of possible complications that may be due to badly managed diabetes:

- Eye complications - glaucoma, cataracts, diabetic retinopathy, and a few others.

- Foot complications - neuropathy, ulcers, and every now and then gangrene which can also require that the foot be amputated

- Pores and skin complications - human beings with diabetes are extra vulnerable to pores and skin infections and pores and skin disorders

- Coronary heart problems - including ischemic heart disease, whilst the blood supply to the coronary heart muscle is diminished
- High blood pressure - not unusual in humans with diabetes, which could enhance the threat of kidney disease, eye troubles, heart assault and stroke
- Intellectual fitness - out of control diabetes raises the risk of laid low with depression, anxiety and a few different intellectual issues
- Hearing loss - diabetes sufferers have a higher danger of developing hearing problems
- Gum sickness - there may be a miles higher prevalence of gum disorder amongst diabetes patients

- Gastroparesis - the muscle groups of the belly prevent running well

- Ketoacidosis - a aggregate of ketosis and acidosis; accumulation of ketone bodies and acidity inside the blood.

- Neuropathy - diabetic neuropathy is a kind of nerve harm that can cause several different issues.

- HHNS (Hyperosmolar Hyperglycemic Nonketotic Syndrome) - blood glucose degrees shoot up too excessive, and there aren't any ketones present inside the blood or urine. It is miles an emergency circumstance.

- Nephropathy - out of control blood strain can result in kidney disorder

- Pad (peripheral arterial disease) - symptoms may additionally consist of pain in the leg, tingling and every now and then issues strolling properly

- Stroke - if blood stress, levels of cholesterol, and blood glucose stages are not managed, the danger of stroke will increase notably erectile disorder - male impotence.

- Infections - people with badly managed diabetes are much more vulnerable to infections.

- Recovery of wounds - cuts and lesions take a great deal longer to heal

Symptoms Caused by Diabetes Complications

It's possible to experience many complications from diabetes that cause others, usually extra drastic and harmful signs. That is why early detection and remedy of diabetes is so critical - it is able to substantially decrease the threat of growing complications like nerve harm, cardiovascular troubles, pores and skin infections, similarly weight advantage/infection and extra.

How in all likelihood are you to experience complication? Numerous elements impact whether or not you will broaden worsened signs or complications because of diabetes, including:

- How well you manage blood sugar stages
- Your blood pressure levels
- How long you've had diabetes
- Your circle of relatives history/genes your
- lifestyle, such as your weight loss program, exercising routine, stress levels and sleep.

Symptoms Related To Nerve Damage (Neuropathy)

A complete half of all people with diabetes will develop a few form of nerve damage, mainly if it is going out of control for decades and blood glucose levels remain abnormal. There are several one of a kind sorts of nerve damage as a result of diabetes that could cause numerous symptoms: peripheral neuropathy (which affects the feet and hands), autonomic neuropathy (which influences organs like the bladder, intestinal tract and genitals), and several other forms that cause damage to the spine, joints, cranial nerves, eyes and blood vessels.

Signs and symptoms of nerve harm due to diabetes can include:

- Tingling within the feet, described as "pins and needles"
- Burning, stabbing or taking pictures pains in the feet and hands

- Sensitive pores and skin that feels very warm or cold
- Muscle aches, weakness and unsteadiness
- Rapid heartbeats
- Trouble sound asleep
- Changes in perspiration
- Erectile disorder, vaginal dryness and lack of orgasms as a result of nerve damage across the genitals
- Carpal tunnel syndrome
- Proneness to accidents or falling
- Changes in the senses, consisting of listening to, sight, taste and odor

- Problem with normal digestion, including frequent belly bloating, constipation, diarrhea, heartburn, nausea, vomiting.

Skin-Related Diabetes Symptoms:

One of the regions affected most and quickest by diabetes is the skin. Diabetes signs on the skin may be some of the most easy to recognize and earliest to show up. Some of the ways that diabetes influences the pores and skin is by causing terrible flow, sluggish wound restoration, reduced immune feature, and itching or dryness. This makes yeast infections, bacterial infections and different skin more easy to increase and harder to do away with.

Skin troubles induced by diabetes consist of:

- Rashes/infections which can be occasionally itchy, hot, swollen, crimson and painful
- Bacterial infections (which includes vaginal yeast infections bacteria, also referred to as staph)
- Styles in the eyes and eyelids
- Acne
- Fungal infections (inclusive of candida symptoms that affect the digestive tract and fungus in pores and skin folds, together with across the nails, below the breasts, between the hands or toes, inside the mouth, and across the genitals) Jock itch, athlete's foot and ringworm

- Dermopathy

- Necrobiosis lipoidica diabeticorum

- Blisters and scales, particularly round infections

- Folliculitis (infections of hair follicles)

Eye-Related Diabetes Symptoms

Having diabetes is one in all the most important factors for developing eye troubles and even imaginative and prescient loss/blindness. Human beings with diabetes have a higher danger of blindness than human beings without diabetes, however most only developed minor problem that can be handled earlier than they get worse.

Diabetes affects the outer, difficult membrane part of the eyes; the front component, which is obvious and curved; the cornea/retina, which consciousness light; and the macula.

Signs and symptoms of diabetes related to imaginative and prescient/eye health can encompass:

- Diabetic retinopathy (a term for all disorders of the retina as a result of diabetes, such as nonproliferative and proliferative retinopathy)
- Nerve damage to the eyes
- Cataracts
- Glaucoma
- Macular degeneration

- Seeing spots, vision loss and even blindness

One of the areas of the eyes maximum impacted by diabetes is the macula, that is specialized for seeing nice details and permitting us to see with sharp vision. Problems with blood drift making its way from the retina to the macula leads to glaucoma, which is 40%more likely to occur in humans with diabetes than in healthy humans. Danger for glaucoma goes up the longer a person has had diabetes and also the older a person turns into.

CHAPTER 7

CONTROLLING DIABETES-TREATMENT IS EFFECTIVE AND IMPORTANT.

All varieties of diabetes are treatable. Diabetes Type 1 lasts an entire life, there may be no known cure. Type 2 normally lasts a lifetime, however, some people have managed to put off their symptoms without medicinal drug, through a combination of exercise, diet and frame weight control.

Gastric bypass surgery can reverse type 2 diabetes in a high percentage of sufferers. Inside three to five years the disease recurs in about 21% of those that did the surgery. "the recurrence rate was greatly stimulated by means of a longstanding records of Type 2 diabetes before the surgical operation. This suggests that early surgical intervention in the overweight, diabetic population will improve the durability of remission of Type 2 diabetes."

Sufferers with Type 1 are treated with regular insulin injections, as well as a special weight-reduction plan and exercise. Sufferers with type 2 diabetes are usually dealt with capsules, workout and a unique diet, but now and again insulin injections are also required. If diabetes is not correctly controlled the patient has a significantly better chance of growing complications.

Natural Ways to Help Control Diabetes Symptoms

Diabetes is a serious condition that incorporates many risks and signs and symptom, however the desirable information is it could be controlled with correct treatment and lifestyle modifications. A high percentage of humans with Type 2 diabetes are able to reverse and manipulate their diabetes signs completely and clearly through enhancing their diets, tiers of bodily interest, sleep and pressure stages. And even though Type 1 diabetes is tougher to deal with and manipulate, complications can also be decreased by taking the same steps. One of the pleasant things to do to prevent diabetes symptoms from worsening is to educate yourself approximately how diabetes forms and worsens.

1. Keep Up with Regular Checkups

Many people with complications of diabetes won't have noticeable symptoms (for example, nonproliferative retinopathy, which may cause vision loss or gestational diabetes at some stage in pregnancy). This makes it certainly crucial that you get looked at by using your doctor frequently to display your blood sugar levels, development, eyes, skin, blood strain tiers, weight and heart.

To be certain that you don't put yourself at a higher risk for coronary heart problem, work with your medical doctor to ensure you hold near normal blood pressure, and triglyceride (lipid) levels. Ideally, your blood pressure shouldn't move over 130/80. You must also attempt to preserve a wholesome weight and decrease inflammation in popular. The quality manner to do that is to devour an unprocessed, wholesome weight loss plan as well as exercising and sleep nicely.

2. Eat a Balanced Diet and Exercise

As part of a healthy diabetes weight loss plan, you could help maintain your blood sugar inside the normal variety via eating unprocessed, complete ingredients and averting things like delivered sugars, trans fats, processed grains and starches, and conventional dairy products. Physical state of being inactive and weight problems(Obesity) are strongly associated with the development of type 2 diabetes, which is why exercise is important to control symptoms and lower the risk for headaches, inclusive of heart ailment. The countrywide institute of fitness states that humans can sharply lower their risk for diabetes through dropping weight via regular bodily pastime and a food plan low in sugar, refined fat and excess energy from processed foods.

3. Control Blood Sugar to Help Stop Nerve Damage

The quality manner to help prevent or delay nerve damage is to closely adjust your blood sugar levels. If you suffer from digestive issues due to nerve damage affects your digestive organs, you may benefit from taking digestive enzymes and supplements like magnesium which could help loosen up muscle groups, improve gut health and control signs. Other problems like hormonal imbalances sexual dysfunctions and trouble dozing also can be significantly decreased when you enhance your eating regimen, nutrient consumption, pressure levels and condition overall.

4. Help Protect and Treat the Skin

Humans with diabetes tend to have extra bacterial, fungal and yeast infections than healthy human beings do. When you have diabetes, you can help prevent pores and skin problems by means of dealing with your blood sugar ranges, practicing proper hygiene and treating skin obviously with things like *vital oils*.

Medical practitioners additionally endorse you limit how frequently you shower whilst your skin is dry, use natural and slight products to clean your skin (in place of many harsh, chemical products sold in most shops), moisturize daily with something mild like *coconut oil for skin*, and avoid burning your skin inside the sun.

5. Safeguard the Eyes

peoples that preserve their blood sugar levels closer to normal are less likely to have vision-related issues or at the least more likely to enjoy milder signs and symptoms. Early detection and appropriate follow-up care can save your Imaginative and prescient. To assist decrease the hazard for eye-related problems like moderate cataracts or glaucoma, you need to have your eyes checked at least one to two instances every year. Staying physically energetic and preserving a healthy weight loss program can save you or delay vision loss by way of controlling blood sugar, plus you need to additionally put on sun shades whilst in the solar. If your eyes become extra damaged over time, your physician may additionally propose you get hold of a lens transplant to hold

imaginative and prescient.

CHAPTER 8

WHAT IS HYPOGLYCEMIA?

Hypoglycemia is a circumstance caused by low blood sugar degrees, additionally on occasion called low glucose. Glucose is mostly found in carbohydrate foods and those containing sugar and is considered to be one of the most crucial sources of energy for the body.

Here's an overview of ways glucose works as soon as it enters the body and the method of ways our hormones adjust blood sugar levels:

- When we consume foods that incorporate glucose (including fruit, veggies, beans, grains and sugary snacks), glucose is absorbed into the bloodstream, in where it eventually carried all through the body into cells for strength.

- In order for our cells to use glucose, the hormone referred to as insulin needs to be present, which is made with the aid of the pancreas in reaction to how a great much glucose we consume.

- Insulin helps our cells absorb the quantity of glucose they need for energy, after which any extra glucose is sent to the liver or various muscle tissues to be stored as glycogen for later use.

- Other than storing glycogen as an energy supply that may be tapped while needed, we can also create fats cells (which shape adipose tissue, or body fat) from greater glucose that we don't need for energy.

- In healthful human beings, when blood glucose levels fall too low, the hormone called glucagon lets the liver understand that it needs to release stored glycogen to keep blood glucose within a healthy range.
- If this technique becomes impaired for any purpose, blood sugar levels remain low and hypoglycemia signs and symptoms develop.

The alternative of hypoglycemia is referred to as hyperglycemia, that's the circumstance as a result of high blood sugar (excessive glucose). Hyperglycemia normally develops in human beings with pre-diabetes or diabetes if their situation isn't well controlled. Hyperglycemia causes signs and symptoms associated with diabetes, consisting of accelerated thirst, urination, fatigue and dizziness.

Diabetics can also experience hypoglycemia if they suffer from drastic fluctuations in blood sugar levels because of mismanagement of insulin and glucose. In people with diabetes, hypoglycemia is often a severe facet impact of taking blood-sugar-lowering medications (containing insulin) that make glucose levels drop too drastically or from not consuming a balanced, healthy diet.

Studies have discovered that repeated episodes of hypoglycemia can negatively impact a person's defense mechanisms towards falling blood glucose, ensuing in giant headaches, together with a six fold increase within the danger of dying from a severe episode.

Hypoglycemia Symptoms and Signs

Ever feel shaky, cranky and tired right before consuming a meal? Or ever dieted and purposefully skipped ingesting, only to crave sugar and sense fatigued? Then you've experienced what it seems like to have low blood sugar. The most common symptoms of hypoglycemia, in other words signs of low blood sugar, include:

- Hunger, every now and then which can be intense and unexpected
- Signs and symptoms of anxiety, which includes nervousness or shakiness
- Sweating, along with night sweats that show up at the same time while sleeping (that is a sign of "nocturnal hypoglycemia")
- Feeling dizzy or light-headed
- Becoming fatigued, worn-out or groggy
- Problem dozing and waking up
- Feeling tired
- Feeling irritable and having mood swings
- Paleness inside the face

- Complications

- Muscle weakness

- Signs and symptoms of brain fog, which include feeling pressured and having problem working or concentrating

- In severe cases (inclusive of when diabetes medicinal drugs are concerned), seizures, coma or even demise can arise. Diabetic patients are at the best risk for severe hypoglycemia episodes, specifically in the event that they happen repeatedly over a long period of time. Severe hypoglycemic episodes in older patients with diabetes had been proven to be associated with an increased risk of dementia, coronary heart ailment, practical brain failure, nerve damage and dying.

Take into account that it's possible to have signs of both hyperglycemia and hypoglycemia whilst blood sugar levels aren't managed. Over the years, these comes with complication and often side effect which might be indicative of pre-diabetes or diabetes, inclusive of fatigue, sugar cravings, modifications in blood stress, weight loss or gain, nerve damage, and anxiousness.

Hypoglycemia Causes and Risk Factors

What are the underlying reasons someone develops hypoglycemia symptoms? The causes of hypoglycemia include:

Mismanagement of insulin

Too much sugar in the blood can cause insulin to rise to high levels, over and over again, which in the long run causes insulin resistance (when cells prevent responding to regular amounts of insulin). This may lead to diabetes or other signs and symptoms of metabolic syndrome in some instances but additionally contributes to fluctuating blood sugar levels in folks that aren't taken into consideration diabetic.

Poor diet

Consuming too little food, going for lengthy intervals without sufficient to devour or having nutrient deficiencies can make a contribution to hypoglycemia. Fad-dieting/crash-dieting plan also can cause symptoms, since these typically involve eating small meals or skipping meals altogether. Some research have found that, usual, insufficient meals consumption was the no. 1 most common cause identified for intense hypoglycemia episodes. Referred to as "impaired counter-regulatory mechanisms," this essentially means that not being attentive to your personal symptoms of starvation can once in a while purpose severe hypoglycemia symptoms.

Diabetes medication

Diabetics are always treated with medication to offset their resistance to insulin's normal effects-in other words to lower high blood sugar. Clinical trials have found that attempt to apply insulin and glucose medication to achieve aggressive healthy blood sugar levels is related to a threefold increase in the risk of hypoglycemia signs. This hypoglycemic effect is now considered by many specialists to be a big problem, even "counter-balancing the benefits of intensive glucose control," according to the Indian journal of endocrinology and metabolism. Medication that can contribute to hypoglycemia consist of chlorpropamide (diabinese), glimepiride (amaryl), glipizide (glucotrol, glucotrol xl), repaglinide (prandin), sitagliptin (januvia) and metformin.

Medication used to treat other disease

When some medications are mixed with insulin, they are able to lower blood sugar too much. These encompass pramlintide (symlin) and exenatide (byetta).

Increased Physical Activity

Over-exercising and overtraining or not eating something after exercise can cause low blood sugar. Muscles deplete glucose inside the blood or stored glycogen to repair themselves, so it's important to refuel after workout routines so that it will prevent symptoms.

Other Health Problem

Hormone imbalances, autoimmune disorders, eating disorders, organ failure or tumors that affect hormone levels can all have an effect on the way insulin is released, glucose is taken up into cells and glycogen is saved.

Alcohol

Alcohol increases blood sugar, but in a while levels can fall too low.

Enzyme deficiencies

Certain metabolic elements could make it difficult to break down glucose properly or for the liver to release glycogen when needed.

High ranges of stress

Pressure can raise cortisol levels, which interferes with how insulin is used.

Conventional Treatment for Hypoglycemia

In line with the American Diabetes Association, conventional treatments for hypoglycemia are usually as follows:

- Making changes to your food regimen and way of life to better control blood glucose. This could encompass converting meal frequency or adopting a diabetic diet plan.

- Doctors often recommend eating 15–20 grams of glucose (from carbohydrates) right away when hypoglycemia symptoms start.

- Hold a watch on signs and symptoms for about 15 minutes, and if you're diabetic, take a look at your blood sugar right now.

- Devour at least a small snack every two to three hours to keep symptoms from returning. Snacks and food have to have at least 15 grams of carbohydrates.

- From time to time doctor prescribe medicines, inclusive of glucose pills or gel, along with other medicines to control hypoglycemia signs and symptoms in diabetic patients. Occasionally injectable glucagon kits are used as a medicine to treat someone with diabetes who has become unconscious from a severe insulin reaction.

Natural Treatments for Hypoglycemia

1. Follow a Hypoglycemia Diet:

If you've had hypoglycemia episodes in the past, try following a balanced meal plan while keeping track of symptoms to discover ways to normalize your blood sugar levels. Foods that can be beneficial for handling hypoglycemia signs and symptoms encompass:

- High-fiber foods: artichokes, green leafy greens, chia seeds, flaxseeds, beans, apples, pumpkin seeds, almonds, avocado and candy potatoes are suitable picks.

- Healthy carbs: carbohydrates are the main nutritional source of glucose, but no longer all carbs are created equal. Exact choices encompass brown or wild rice, candy potatoes, sprouted ancient grains, legumes, and beans.

- Vegetables and entire portions of fruit: fruit and clean fruit juice can be particularly helpful to offset a hypoglycemic episode.

- Healthful fat: virgin coconut oil, MCT oil, extra virgin olive oil, nuts and seeds (like almonds, chia, hemp and flax), and avocado are appropriate sources.

- Quality protein: wild fish, which include salmon, free-range eggs, grass-fed red meat or lamb, uncooked dairy products (which includes yogurt, kefir or raw cheeses), and pasture-raised rooster are some of the best protein foods.

Foods that should be avoided encompass:

- Too much caffeine or alcohol
- Empty calories, consisting of packaged goods that are incredibly processed
- Lots of added sugar
- Sweetened drinks

- Refined grains
- Fast food and fried foods

2. Rethink skipping foods or cutting calories too low.

Peoples with hypoglycemia or diabetes have to devour regular meals throughout the day, have sufficient calories at each meal (typically inclusive of some healthy carbohydrates) and by no means skip meals altogether. Wholesome snacks every few hours can also be beneficial for keeping blood sugar strong and preventing dips in strength.

If you're exercising and sense week or dizzy, make sure you're consuming enough, take a break and consider having something small to consume beforehand. Refuel after exercises with a snack that consists of a combination of protein and healthy carbs. If you notice that you have symptoms of hypoglycemia during the night while sleeping, remember having a snack before bed to prevent hypoglycemia overnight.

3. Speak to your health practitioner about your medication

In case you take any medication that alter blood glucose or insulin levels, be very cautious to reveal physical signs and symptoms cautiously that could point to hypoglycemia. Research suggests that signs and symptoms of hypoglycemia can end up regularly less intense over time or maybe diminish altogether, resulting in "hypoglycemia unawareness" in a significant proportion of patients with repeated episodes because of medicinal drugs. Talk to your medical doctor approximately how you may track your blood sugar levels more appropriately or in case your dosage must be changed to lower signs and symptoms.

Facts about Hypoglycemia

- Restricting calorie intake (via dieting, fasting or skipping meals) has been diagnosed as the no. 1 reason of hypoglycemic episodes. Different main causes encompass an excessive amount of exercise without refueling and taking unhealthy doses of insulin medications.

- Insulin medication can sometimes trigger severe hypoglycemia episodes, even ones that may be deadly. Evidence from several studies suggests that severe hypoglycemia occurs in 35 percentage to forty-two percent of diabetic patients taking insulin medicines, and the common charge of severe hypoglycemia assaults is among 90-130 episodes over the path of a patient's lifetime.

- Studies have located that the longer someone has diabetes (as an example, over 15 years), the higher their risks becomes of having repeated episodes of severe hypoglycemia symptoms.

- In Type 1 diabetic patients who have no longer been diagnosed or treated, threat for death is significantly higher than in healthy individuals. For example, nocturnal hypoglycemia bills for 5 percent to 6 percent of all deaths amongst young people with Type 1 diabetes.

- Inside the U.S., the estimated number of emergency department visits because of hypoglycemia is about 298,000 per year.

- To help prevent hypoglycemia signs and symptoms, most people should consume something each 3 to 4 hours and attempt to devour at least 15 grams of carbs with every meal.

Hypoglycemia blood sugar chart:

Wondering what stages of glucose within the blood are considered to be too high or too low? Generally speaking, experts agree that are no clear-cut between the normal range of blood sugar and high and coffee blood sugar levels. However, researchers and medical doctors regularly use the subsequent blood sugar chart to categorize different condition

Normal blood sugar:

Around 60-140 milligrams of sugar according to deciliter of blood (mg/dl) is taken into consideration to be inside the range of healthy blood sugar. There is a normal "range" because even completely healthy people experience some fluctuations in blood sugar ranges for the duration of the day relying on how they consume or their levels of activities. The international unit for categorizing Healthful blood glucose is 3.3 and 7.8millimole according to liter (mmol/l).

If you're generally healthy (you don't have diabetes) and also you haven't eaten anything in the past eight hours (you've been "fasting"), it's normal for blood sugar to be anything among 70-99mg/dl (less than 100mg/dl). In case you're healthy and also you've eaten inside the past two hours, it's normal for blood sugar to be anything less than a 140mg/dl.

Hypoglycemia

Usually taken into consideration whatever beneath 60-70mg/dl. If you do have a history of diabetes, fasting glucose must preferably also be under 100mg/dl, which would possibly want to be managed through using insulin. It's also consider healthy to have levels between 70–130 prior to eating. When you have diabetes, you want to keep blood sugar between 100-140mg/dl prior to bedtime and at the least a 100mg/dl prior to exercise.

Hyperglycemia

If Type 1 diabetes is left untreated, sometimes blood glucose can rise to 500 mg/dl (27.8 mmol/l). Levels this high are rarer in people with Type 2 diabetes; especially they take medication or use a healthy way of life to monitor their stages. When you have diabetes and you've eaten in the past two hours, the intention is to have blood sugar stay under 180 mg/dl.

Precautions Regarding Hypoglycemia

Always go to a medical doctor or the emergency room if you observe intense and sudden symptoms of hypoglycemia, including fainting. If you ever turn out to be unconsciousness or have a seizure and additionally take medicines that might adjust blood glucose, surely mention this to your medical doctor.

If you're diabetic, it's recommended that you train someone how to administer glucagon to treat severe hypoglycemic events and have that man or woman call 911 if there's an emergency right away. Don't forget about severe symptoms, which include passing out, insomnia, fast coronary heart beats, and many others., that continue over time, as this increases the chance for lengthy-term complications.

Final Thoughts on Hypoglycemia

- Hypoglycemia is a condition characterized by means of abnormally low blood glucose (blood sugar) levels.

- Common signs and symptoms of hypoglycemia consist of hunger pangs, shakiness, irritability, dizziness and fatigue.

- Reasons of hypoglycemia consist of cutting calories, skipping meals, a poor weight loss program, nutrient deficiencies and no longer ingesting after exercise.

- Severe hypoglycemia signs and symptoms have an effect on humans with diabetes who're taking medications most usually and are from time to time known as insulin reaction or insulin shock.

- Natural treatment for hypoglycemia signs and symptoms encompass consuming frequently every few hours, eating a balanced diet, refueling after exercise and being careful not to overdue medications that interfere with blood sugar regulation.

CHAPTER 9

WHAT IS HYPERGLYCEMIA?

Hyperglycemia is a term referring to high blood glucose levels - the conditions that regularly warrant diagnosis of diabetes.

High blood glucose levels are the defining feature of diabetes, but once the disease is recognized, hyperglycemia is a signal of poor control over the condition. Hyperglycemia is described with the aid of positive high levels of blood glucose:

- Fasting ranges more than 7.0 mmol/l (126 mg/dl)
- Two-hour postprandial (after a meal) levels greater than 11.0 mmol/l (200 mg/dl).

Chronic hyperglycemia typically results in the development of diabetic complications.

Symptoms of Hyperglycemia

The most common symptoms of diabetes itself are related to hyperglycemia - the traditional symptoms of frequent urination and Thirst. Typical signs and symptoms of hyperglycemia that has been showed by way of blood glucose measurement include:

- Thirst and starvation
- Dry mouth
- Common urination, especially at night
- Tiredness

- Recurrent infections, together with thrush

- Weight loss

- Imaginative and prescient blurring (vision blurring).

Hyperglycemia is defined as having a fasting blood glucose level more than 126 mg/dl

Causes of Hyperglycemia

Hyperglycemia often ends in the analysis of diabetes. For people already diagnosed and handled for diabetes, but, poor control over blood sugar levels results in the condition. Causes of this include:

- Consuming more or exercising less than normal

- Insufficient quantity of insulin treatment (more commonly in case of Type 1 diabetes)
- Insulin resistance in type 2 diabetes
- Contamination such as the flu
- Mental and emotional stress
- The "dawn phenomenon" or "dawn effect" - an early morning hormone surge.

Treatment and Prevention Of Hyperglycemia

Prevention of hyperglycemia for human beings with a diabetes diagnosis is a matter of appropriate self-tracking and management of blood glucose levels, consisting of adherence to insulin regimes if essential.

For someone who has not been diagnosed with diabetes, symptoms of hyperglycemia need to be reported to a medical doctor who will test for diabetes - other conditions also can cause hyperglycemia.

Control of high blood sugar is vital to prevent complications caused by chronic hyperglycemia. A medical doctor may need to check the treatment plan for a diabetes patient who becomes hyperglycemic and they'll decide to take one of the following actions:

- Raise the insulin dose
- Recommend dietary changes

- Recommend more exercise

- Endorse closer glucose tracking

Hyperglycemia Can Lead To

Diabetic Ketoacidosis

It is miles crucial to attend to hyperglycemia due to the fact it could result in a dangerous complication referred to as ketoacidosis that could result in coma or even death. Ketoacidosis not often occurs in Type 2 diabetes, commonly occurring in cases of Type1 diabetes. High levels of glucose within the blood imply that insufficient levels of glucose are available to cells for their energy needs. As a result, the body resorts to breaking down fats so that strength is derived from fatty acids. This breakdown produces ketones, leading to higher acidity of the blood.

Diabetic ketoacidosis requires urgent clinical attention and, along hyperglycemia and its signs, is signaled by:

- Nausea or vomiting
- Stomach pain
- A fruity smell on the breath
- Drowsiness or confusion
- Hyperventilation
- Dehydration
- Lack of consciousness.

Medical treatment of ketoacidosis consists of the administering of intravenous fluids and insulin.

CHAPTER 10

WHAT IS INSULIN?

Insulin is a hormone; a chemical messenger produced in one a part of the body to have an action on some other. It is a protein responsible for regulating blood glucose levels as a part of metabolism.

The body manufactures insulin in the pancreas, and the hormone is secreted via its beta cells, basically in response to glucose.

The beta cells of the pancreas are flawlessly designed "fuel sensors" stimulated with the aid of glucose.

As glucose levels increases in the plasma of the blood,

uptake and metabolism by using the pancreas beta cells are more suitable, leading to insulin secretion.

Insulin has two modes of action on the body - an excitatory one and an inhibitory one:

- Insulin stimulates glucose uptake and lipid synthesis

- It inhibits the breakdown of lipids, proteins and glycogen, and inhibits the glucose pathway (gluconeogenesis) and production of ketone bodies (ketogenesis).

Taking Insulin

Having insulin-dependent diabetes means a lifelong dependence on daily injections of insulin. In addition to People with type 1 diabetes, people with type 2 diabetes that is unresponsive to oral tablets must also take insulin.

An ordinary affected person with type 1 diabetes may need more than 60,000 injections throughout their lifetime, requiring two or more injections every day.

"ever since the creation of insulin for the remedy of diabetes, methods of administering it other than with the aid of injection have been investigated." that is a quote from a paper posted within the lancet in 1940, and investigations continues to the present time.

The main impediment to finding a way of delivering insulin in pill form is the digestive system itself - either the gut breaks the insulin down or the insulin moves through intact due to the fact that it is unable to pass via the gastrointestinal membrane.

Different Preparations of Insulin

In the US, all insulin that is offered has been manufactured in a laboratory. Even though animal insulin used to be available, Preparations from pigs and livestock have now been withdrawn from the market. Information and warnings about importing these animal preparations from abroad have been published with the aid of the US Food and Drug Administration.

Analogs of human insulin are manufactured forms with some structural modifications built in, differing in their amino acid Series to alter their pharmacological characteristics.

Insulin may be manufactured to produce different actions. The speedy-action insulin are insulin glulisine, insulin lispro and insulin aspart. Short-performing insulin are insulin regular, while intermediate-performing insulins are impartial protamine hagedorn (NPH) insulin, also referred to as isophane insulin. Subsequently, the long-acting insulins are insulin detemir and insulin glargine.

Different preparation of insulin provide a wide range of options in terms of ways quickly they take effect, their peak time of action and their all in all duration of effect:

- Speedy-acting insulin analogs have an onset of action at between 5-15 minutes, a peak action at 30-90 minutes and an overall duration of effect of 3-5 hour

- Short-acting regular insulin has an onset of action at among 30-60 minutes, a peak motion at 2-3 hours and a universal period of impact of 5-8 hours. The foremost time for injecting is 30 minutes before eating.

- Intermediate-acting insulins have an onset of action at between 2-4 hours, a peak action at 4-12 hours and an overall duration of impact of 10-18 hours

- Lengthy-acting insulins have an onset of action at among 2-10 hours, a peak action at 6-16 hours (except insulin glargine, which has no peak) and an universal duration of effect of 16-24 hours. These insulins maintain glucose levels fairly uniformly over a 24-hour length.

Insulins may also be combined at 30:70, 25:75 and 50:50 combinations to produce two peak times of action.

Practical Advice for Injecting Insulin

With practice and precise method, injecting insulin can turn out to be more comfortable. The needle may be very small, and injection is not right into a muscle or vein but beneath the skin. The three regions of skin most generally used are the stomach, the buttocks and the thighs.

The choice of site depends on a number of factors however can be circled to help avoid the formation of lumps. Different sites depend in different rate of absorption. Insulin is absorbed quickest through the abdomen, followed by the arm, after which the thigh and finally the buttocks. If physical exertion takes place after injection, this also increases absorption by way of increasing blood flow. Rubdown of the injection site also has an effect.

Injections to the identical place should be varied by way of preserving injections a couple of finger widths apart. Other realistic tips include:

- Avoiding the belly button, the inner thigh, the lower buttock, scars and damaged blood vessels or varicose veins
- If using the thigh, keep injections at least four inches under the pinnacle of the leg and above the knee
- If using the arm, inject into the fatty region at the back, between the shoulder and elbow

- If you are using the buttock, make sure that you use the hip area.

Insulin Side Effects

Not quite often, an extreme and life-threatening hypersensitivity may be experienced after insulin injection. This anaphylaxis is a clinical emergency requiring instant medical care. Severe insulin side effects and anaphylactic reactions are signaled with the aid of:

- Rash or itching over the whole body
- Swelling (edema) of the tongue, throat, hands, hands, feet, ankles or lower legs
- Problem/difficulty in breathing or shortness of breath

- Issue swallowing
- Wheezing
- Dizziness
- Blurred vision
- Fast heartbeat (tachycardia) or abnormal heartbeat rhythm
- Sweating
- Weak spot
- Muscle cramps
- Significant weight gain in a short period of time.

Progressively increasing insulin doses under medical supervision is used as a treatment to desensitize an individual with a severe insulin allergy. Side effects of insulin that are more common include:

- Hypoglycemia - low blood sugar levels which can result from the timing of the insulin injection. Hypoglycemia can probably be avoided by using pre-dinner dose of intermediate-acting insulin to bedtime dose

- Weight gain - this can manifest initially while insulin therapy has begun, due to correction of protein and power metabolism. Later weight gain may be caused by fluid retention or immoderate eating due to hypoglycemia

- Lipohypertrophy - raised lumps in the skin as a result of repeated injections at the same site; this can be avoided by means of rotating the injection sites

- Other local effects- those are less common than lipohypertrophy and consist of infection, injection site abscess (both of which can be avoided with good injection practices), allergic reaction and lipoatrophy (loss of fat tissue).

Inhaled Insulin

Human insulin inhalation powder (Afrezza) became available by prescription inside the US in February 2015, the handiest inhaled insulin to be had at the time. It is available at around twice the cost of the injected rapid-acting insulin.

Afrezza is a rapid acting, dry-powder formulation of recombinant human insulin manufactured by mankind and sanofi and can be used in the treatment of the adults with type 1 or type 2 diabetes. In sufferers (patients) with Type 1 diabetes, the drug should be used in combination with lengthy-acting insulin. A single inhalation of Afrezza is taken at the beginning of meals.

This is not the first inhaled insulin product to reach the marketplace-a rapid-acting insulin, exubera, was approved in 2006 but withdrawn just a year later via its manufacturer. That early device was a bulky size- approximately the dimensions (size) of a flashlight.

Afrezza, however, is introduced through a smaller, palm-sized tool that can be held without problems between thumb and finger.

Long-term evidence on the safety of Afrezza is still to be amassed. To this present moment, coughing has been identified as a common side effect, and there has been evidence of throat pain and irritation. As with other insulin, it can also lead to hypoglycemia.

Patients with chronic lung disease inclusive of bronchial asthma and chronic obstructive pulmonary disease (COPD) are contraindicated from using Afrezza because it increases the risk of bronchospasm. The formulation must be averted by the patients who smoke or have stopped smoking within the last 6 months.

Inhaled insulin appears to be similarly effective to injected insulin at controlling blood glucose levels. One review says that Afrezza have to be reserved for any other healthy adults with diabetes who do not have lung disorder/disease and who are unwilling or not able to use injectable insulin.

CHAPTER 11

INSULIN PUMP FOR CONTROLLING DIABETES.

Ever since the discovery of insulin and its use in treating diabetes, medical research has struggled to find a way of delivering it that accurately mimics the normal physiological action of insulin and overcomes the burden of daily injections.

The main development in this area has been the insulin pump. Researchers are also aiming to develop a fully automated artificial pancreas. Other means of delivering insulin have been launched and continue to be investigated.

Insulin Pumps for Type 1 Diabetes.

Insulin pumps - or continuous subcutaneous insulin infusion pumps - cast off the daily need for a couple of injections. Instead, a cannula - a completely skinny and flexible plastic tube inserted under the pores and skin using a needle - needs to be replaced every two or three days. As well as requiring fewer needles, insulin pumps may be appealing due to offering flexibility in meal timing and low variability in glucose levels.

Insulin pumps also have its own disadvantages, despite the fact that for most of the users these are outweighed by using the benefits. Downsides encompass:

- Higher cost, with a few but not all insurance carriers covering their expenses
- The inconvenience of wearing an external device - sores may develop at the needle site
- Training is required - frequent and cautious self-monitoring of glucose levels is needed for secure and powerful use, as is a sound understanding of the pump's function
- Mechanical failure could occur, ensuing in interruptions to insulin supply.

An insulin pump continuously releases insulin in small doses (the basal insulin) from its reservoir and may deliver an extra dose (a bolus) when needed. As a result, an insulin pump more closely mimics normal insulin physiology and offers greater accuracy than day by day injections. Insulin pumps also can provide better glucose control and better HbA1c readings.

In addition to the greater flexibility afforded in meal planning, using an insulin pump cuts out the unpredictable effects or intermediate- or long-acting insulin, as well as allowing for exercise without the need for high carbohydrate consumption.

However, insulin pumps also can cause weight gain and lead to complication if the catheter comes out, resulting in missed Insulin.

INSULIN PUMPS FOR TYPE 2 DIABETES

Type 2 diabetes has seen more use of external insulin pumps compared with their set up use in cases of type 1 Diabetes. Using insulin pumps in cases of Type 2 disease is subject to debate and there is less evidence supporting their use.

In a controlled trial of insulin pump versus daily injection, however, researchers publishing in the lancet in 2014 concluded:

In patients with poorly controlled type 2 diabetes despite using a couple of daily injections of insulin, pump treatment may be taken into consideration as a save and valuable treatment option."

Future Options for Delivering

Insulin

Studies into the use of two hormones in pumps, insulin and glucagon - bi-hormonal therapy - are ongoing. The stableness of glucagon, however, remains an impediment to success.

The goal of this sort of combination is to counter the excess effects of insulin with glucagon - a function that this hormone fulfills in peoples without diabetes. The concept has also shown some promise in lowering the risks of hypoglycemia.

Commercially available glucagon is not stable in aqueous solution for lengthy durations, however, forming potentially cytotoxic fibrils that accumulate speedy and might turn to be a gel with the capacity to impede the pump. Researchers are currently searching for a way to find solution to this problem.

Artificial Pancreas

The idea of changing the role of the pancreas in monitoring and regulating glucose levels has been pursued since the 1960's.

Referred to as the artificial pancreas, the idea is to have a "closed-loop control" of blood glucose in diabetes with a system that combines:

- A glucose sensor to measure blood glucose levels
- Computer controllers that use a mathematical model of the metabolic gadget to calculate insulin doses
- An insulin infusion device similar to an insulin pump.

A wearable synthetic pancreas has become a feasible and secure prospect in current years, but research continues.

CHAPTER 12

WHAT IS ORAL INSULIN?

Perhaps the best way to deliver insulin would be via the mouth in pill (tablet) form, the way and manner most daily medication is taken.

Unfortunately, insulin puts up fundamental obstacles to the development of an oral form:

- The gastrointestinal tract's proteolytic enzymes break insulin down
- Insulin does not does not transport readily across the gastrointestinal membrane.

In spite of these obstacles, several research studies have achieved some positive results. Some delivery systems are even in advanced stages of improvement.

As soon as these practical hurdles have been overcome, it pays to be demonstrated whether or not oral delivery will be much more secure or more powerful than needle delivery. At present, the economic reality of oral insulin remains elusive.

Other potential forms of insulin delivery require further research. This includes transmucosal delivery-through an Intranasal or oral spray, for example-and transdermal delivery using a skin patch.

Treatments for Diabetes

Type 2 diabetes has a number of drug treatment options to be taken by means of mouth referred to as oral anti-hyperglycemic drugs or oral hypoglycemic drugs.

Oral diabetes tablets are generally reserved to be used most effective after lifestyle measures were unsuccessful in lowering glucose levels to the target of an HbA1c under 7.0%, achieved through an average glucose reading of around 8.3-8.9 mmol/l (round 150-160 mg/dl).

The lifestyle measures which can be critical to type 2 diabetes control are weight loss plan (diet) and exercising, and these remain an crucial part of treatment whilst pills are added. People with Type 1 diabetes cannot use oral pills for treatment, and can take insulin as the best and possible option.

How do oral drugs lower glucose levels?

Oral anti-hyperglycemic drugs have 3 modes of action to reduce blood glucose levels

- Secretagogues stimulate insulin secretion by the pancreas
- Sensitizers increases the sensitivity of the peripheral tissues to insulin
- Inhibitors impair gastrointestinal absorption of glucose.

Each class of anti-hyperglycemic drugs has an exceptional negative effect or safety profile, and side consequences are the principle consideration when it comes to choosing a medicinal drug. Possible side effect range from weight gain, via gastrointestinal ones consisting of diarrhea, to pancreatitis and more serious problems. Hypoglycemia is likewise a likely detrimental event.

What oral drugs are available for type 2 diabetes?

No person precise choice of oral hypoglycemic is considered the handiest form of treatment-the decision over which drug to use is instead based on:

- Consideration of the negative aspect effects
- Convenience and average tolerability
- Non-public (personal) preference.

In fact, weighing up each drug is something to do in partnership with a prescriber-Guidelines partly drawn up by the America Diabetes Association list a splendid variety of benefits and downsides for every of the available drug treatments, which include the consideration of price.

The use of a single drug may be escalated to mixture therapy with a second drug in an effort to enhance glycemic control.

Metformin is commonly the first treatment offered, but, and it is the most widely used oral anti-hyperglycemic. Metformin is a sensitizer within the magnificence referred to as biguanides; it really works by lowering the amount of glucose released with the aid of the liver into the bloodstream and increasing cellular response to insulin. A metformin tablet is generally taken twice a day.

This drug is a low-cost anti-hyperglycemic with slight side effect that can encompass diarrhea and abdominal cramping. Metformin isn't always associated with weight gain or hypoglycemia.

Sulphonylureas are secretagogues that increases pancreatic insulin secretion. There are several drug names in this class, inclusive of:

- Chlorpropamide
- Glimepiride
- Glipizide
- Glyburide.

Again, the selection of drug is an individual one. In the case of sulphonylureas, the choice relies upon on daily dosing and the extent of side effects. These drugs are associated with weight gain and hypoglycemia.

Glitazones (also known as thiazolidinediones) are sensitizers-they increase the effect of insulin inside the muscle and fats and reduce glucose production by the liver.

Two type of glitazones are available: pioglitazone and rosiglitazone. These drugs may have the side effects of weight benefit or swelling and are associated with expanded risks of coronary heart problem and stroke, bladder cancer and fractures.

Alpha-glucosidase inhibitors are intestinal enzyme inhibitors that block the breakdown of carbohydrates into glucose, lowering the quantity absorbed within the gut.

To be had as acarbose and miglitol, they're not typically attempted as first-line tablets because of common side effect of flatulence, diarrhea and bloating, despite the fact that those may reduce as the years goes by.

Dipeptidyl peptidase-4 (DPP4) inhibitors encompass alogliptin, linagliptin, saxagliptin and sitagliptin.

Also referred to as gliptins, DPP4 inhibitors have a number of effects, including stimulating pancreatic insulin (by means of preventing the breakdown of the hormone GLP-1). They will also help with weight reduction via an effect on urge for food (i.e appetite).

These drugs do not increase the risk of hypoglycemia. Mild possible side effects are nausea and vomiting.

Sodium-glucose co-transporter 2 (SGLT2) inhibitors include canagliflozin and dapagliflozin. They work by way of inhibiting the reabsorption of glucose in the kidneys, causing glucose to be excreted inside the urine (glycosuria).

SGLT2s might also purpose modest weight reduction. Side effects consist of urinary infection.

Meglitinides encompass repaglinide and nateglinide. They stimulate the release of insulin via the pancreas. Meglitinides are associated with a higher chance of hypoglycemia and need to be taken with meals 3 times a day. As a result, these drugs are less commonly used.

CHAPTER 13

5-STEP PLAN TO REVERSE DIABETES

Step 1: Remove These Foods to Reverse Diabetes Naturally.

Certain meals negatively affect your blood sugar levels, Causes inflammation and trigger immune responses. To reverse diabetes naturally, step one is to put off these foods from your weight-reduction plan:

1. **Refined sugar**: Refined sugar swiftly spikes blood glucose, and soda, fruit juice and different sugary liquids are the worst culprits. These types of sugar enter the bloodstream unexpectedly and might cause severe

elevations in blood glucose. Despite the fact that natural sweeteners like uncooked honey and maple syrup are better options, they can still have an effect on blood sugar level, so it is best to use these foods on occasion. Your first-rate alternative is to switch to **stevia**, a natural sweetener that won't have much effect.

2. **Grains:** grains, in particular gluten-containing grains like wheat, contain massive amounts of carbohydrates that are broken down into sugar within only a few minutes of consumption. Gluten can cause intestinal infection/inflammation, which affects hormones like *cortisol* and *leptin*, and may lead to spikes in blood sugar. I suggest removing all grains out of your diet for 90 days

as your body adjusts to this recuperation program. Then you may attempt bringing sprouted ancient grains returned into your diet program in small amounts.

3. **Conventional cow's milk:** Conventional cow's milk and dairy merchandise must be removed, in particular for humans with Type 1 diabetes. Dairy can be a wonderful food for balancing blood sugar if it comes from goat's, sheep or A2 cows. However live away from all other varieties of dairy due to the fact the A1 casein produced by means of conventional cows will damage the body and trigger an immune response just like gluten. While buying dairy, only purchase uncooked and organic products from pasture-raised animals.

4. **Alcohol:** alcohol can dangerously increase blood sugar and lead to liver toxicity. Studies published in Annals of internal medicine discovered that there was a 43% increased occurrence of diabetes associated with heavy intake of alcohol, which is described as three or more drinks per day. Beer and sweet liquors are especially high in carbohydrates and ought to be avoided.

5. **GMO Meals:** GMO corn, soy and canola had been linked to kidney and liver sickness and can promote diabetes. I advocate getting rid of all GMO foods and all packaged foods out of your diet. Choose products which might be categorized organic or GMO-free.

6. **Hydrogenated oils:** do away with hydrogenated, rancid oils from your diet, which include vegetable oil, soybean oil, cottonseed oil and canola oil. Because these oils are processed, treated with at very high temperatures, and blended with bleaching marketers and artificial dyes, consuming them has been linked to many fitness worries, which include diabetes.

Step 2: Incorporate These Foods to Treat Diabetes

To reverse or prevent type 2 diabetes, add the following foods into your diet:

1. **Foods high in fiber:** research indicates that ninety percent of the U.S. Population doesn't eat sufficient fiber on a day basis. High-fiber foods help slow down glucose absorption alter your blood sugar levels and guide detoxing. Aim to devour at least 30 grams of fiber per day, which could come from veggies (like Brussels sprouts, peas and artichokes), Avocados, berries, nuts and seeds, particularly chia seeds and flaxseeds.

2. **Foods high in chromium:** chromium is a nutrient that's involved in regular carbohydrate and lipid metabolism. Foods high in chromium can improve the glucose tolerance issue to your body and evidently balance out blood glucose levels. It plays a role in insulin

pathways, helping bring glucose into the cells so that it can be used for bodily energy. Broccoli has the best amounts of chromium, but you can also find it in raw cheese, green beans, brewer's yeast and grass-fed beef.

3. **Magnesium-wealthy meals:** magnesium can help regulate blood sugar levels because it performs a role in glucose metabolism. Research shows that diabetes is frequently related to magnesium deficiency. Consuming magnesium-rich foods , like spinach, chard, pumpkin seeds, almonds, yogurt and black beans, can enhance Type 2 diabetes symptoms.

4. **Healthy Fats:** medium-chained fatty acids found in coconut and pink palm oil can help balance blood sugar levels, and they serve as the preferred fuel source to your body rather than sugar. Using coconut milk, ghee and grass-fed butter also can help balance out your blood sugar levels, so include these foods into your meals and smoothies.

5. **Clean protein:** Eating **protein foods** has a minimum effect to your blood glucose levels, and it can sluggish down the absorption of sugar. Some of the best sources of clean protein include wild-caught fish, which includes omega-3 fat that lessen inflammation, grass-fed beef, organic chicken, lentils, eggs and bone broth.

6. **Foods with a low glycemic load:** the glycemic index of a food tells you about the blood glucose raising capacity of the meals. Foods which have a high glycemic index are transformed into sugar after being eaten more quickly than low glycemic Foods. If you are fighting diabetes, stick to low glycemic meals like non-starchy veggies, stone fruits and berries, nuts, seeds, avocados, coconut, natural meat, eggs, wild-caught fish, and raw pastured dairy.

Step 3: Take These Supplements for Diabetes

1. **Chromium Picolinate:** Taking 200 micrograms of *chromium picolinate* three times every day with meals

can help improve insulin sensitivity. A review published in diabetes technology and therapeutics evaluated thirteen studies that reported significant development in glycemic control and substantial reduction in hyperglycemia and hyperinsulinemia after patients used chromium picolinate supplementation. Other positive outcome from supplementing with chromium picolinate blanketed reduced cholesterol and triglyceride levels and decreased requirements for hypoglycemic medicinal drug.

2. **Cinnamon:** Cinnamon has the potential to decrease blood sugar ranges and enhance your sensitivity to insulin. A study carried out at Western University of

Health Sciences in pomona, calif. Found that the consumption of cinnamon is related to a statistically enormous decrease in plasma glucose levels, LDL cholesterol and triglyceride levels. Cinnamon intake also helped increase HDL cholesterol levels. To take advantage of the various health benefits of cinnamon, add one teaspoon to meals, smoothies or tea. You can also take one to two drops of cinnamon essential oil internally by adding it to food or tea, or combine three drops of cinnamon oil with half of a teaspoon of coconut oil and rub it into your wrists and stomach/abdomen.

3. **Fish oil:** Taking a fish oil supplement can help enhance markers of diabetes by reducing triglyceride

levels and raising HDL levels of cholesterol. Research published in the journal of Research in Medical Sciences suggests that omega-3 fatty acids found in fish oil are essential for insulin to function properly, stopping insulin intolerance and reducing inflammation. To use fish oil as a natural treatment for diabetes, take 1,000 milligrams each day.

4. **Alpha Lipoic Acid:** Alpha lipoic acid is an antioxidant that helps turn glucose into gasoline for the body. It correctly Improves insulin sensitivity and decreases symptoms of diabetic neuropathy, including weakness, pain and numbness that's caused by nerve damage. Despite the fact that we make alpha lipoic acid and it is able to be determined in a few food assets, like

broccoli, spinach and tomatoes, taking an ALA supplement will increase the amount that circulates to your body, which can be extremely useful when trying to reverse diabetes naturally.

5. **Bitter melon extract:** bitter melon help decrease blood glucose levels, and it regulates the body's use of insulin. Research shows that bitter melon extract can help lessen and manage symptoms of diabetes, inclusive of insulin resistance, heart complication, kidney damage, blood vessel damage, eye disorders and hormone irregularities.

Step 4: Follow This Eating Plan to Reverse

Diabetes

In case you want to balance your blood sugar and see outcomes fast, then observe this diabetes eating plan as intently as feasible. Consciousness on getting masses of clean protein, wholesome fats and fiber into each meal that can be of help to reverse diabetes.

Review the list of meals that you must be ingesting from step 2, and bring those healthy, diabetes-fighting foods into your diet plan as well. It could seem like a major change in your eating regimen in the beginning, but after a while you will start to observe the wonderful effects these meals are having on your body.

DAY 1

BREAKFAST: Coconut Smoothie 4oz coconut milk, 3 tbsp chia or flax seeds, 1 scorp organic powder, 1tsp cinnamon, Stevia to taste.

LUNCH: Large Spinach Chicken salad, Dressing apple cider vinegar and olive oil.

SNACKS: ¼ Cup Row Almonds.

DINNER: Grass-fed Beef (no bun) with steamed broccoli

DAY 2

BRAEKFAST: Vegetable Omelet with goat cheese cooked in coconut oil.

LUNCH: Chicken vegetable Soup made with real bone broth.

SNACKS: 2oz Raw cheese (A2 Only)

DINNER: Wild-Caught Salmon grilled with spinach and onions sautéed in coconut oil.

DAY 3

BREAKFAST: Peach Probiotic Shake, 4oz goat's milk 2 tbsp almond butter, ½ cup frozen peaches, vanilla protein powder, 1 tsp cinnamon

LUNCH: Turkey Burger with small salad

SNACKS: Raw veggies with guacamole

DINNER: Chicken Vegetable Stir-fry

Some recipes that fit into this eating plan consist of:

- Eggs benedict recipe

- Turkey-filled bell peppers

- Cucumber salad recipe

- Buffalo cauliflower

Step 5: Exercise to Balance Blood Sugar

Exercise reduces chronic disease and can help reverse diabetes naturally. Studies show that Exercise improves blood glucose control and may save you or delay type 2 diabetes, while it can definitely affects your blood pressure, heart health, Cholesterol levels and quality of life.

Exercising naturally helps your metabolism with the aid of burning fats and building lean muscle. To prevent and reverse diabetes, make exercise part of your daily routine. This doesn't suggest that you need to spend time at the gym. Simple varieties of bodily activity, like getting outside and strolling for 20 to 30 minute every day, can be extraordinarily beneficial, especially after food. Practicing yoga or stretching at domestic or in a studio is another superb alternative.

In addition to walking and stretching exercises, try interval training cardio, like burst training, or weight education for 3 to 5 days per week for 20-40 minutes. Burst training helps you to burn up to a few times more body fat than traditional cardio and can naturally increase insulin sensitivity. You may do that on a spin motorcycle with intervals, or you can attempt burst training at home.

Strength training using free weights or machines is likewise encouraged because it allows you build and maintain muscle, which supports balanced blood sugar and sugar metabolism.

Final Thoughts on How to Reverse Diabetes

- More than 30 million human beings-almost 10 percent of the population-have diabetes within the U.S., including the children. Nearly 7.2 million of these people aren't even aware of it.
- Another 34 million adults are pre-diabetic.
- It is the seventh leading cause of death inside the U.S. and costs hundreds of billions of dollars per year.

- Type 1 diabetes typically takes place before someone turns 20 and is not often reversed, but it could be managed via weight loss plan and lifestyle modifications.
- Type 2 diabetes is more common and commonly takes place in people over forty, especially if they're overweight.
- A good way to reverse diabetes naturally, remove foods like refined sugar, grains, traditional cow's milk, alcohol, GMO foods and hydrogenated oils out of your weight loss plan (diet); incorporate wholesome foods like foods high in fiber, chromium, magnesium, healthful fat and clean protein, along with foods with low glycemic loads;

take dietary supplements for diabetes; observe my diabetic eating plan; and exercise to balance blood sugar.

CHAPTER 14

SELF-MONITORING OF BLOOD GLUCOSE

Tight control of blood sugar levels is difficult to attain. Levels can fall too low despite the nice adherence to demanding day by day self-monitoring schedules.

The proportion of human beings inside the US with a diagnosis of diabetes who undertakes self-monitoring of glucose has risen dramatically - from 36% in 1994 to 64% in 2010.

All patients newly recognized with type 1 diabetes will obtain training on how to do their blood sampling and how to act on readings. Increasing numbers of peoples with type 2 diabetes - even those who do not need insulin treatment - also are encouraged to self-monitor their blood glucose levels.

What Is Blood Glucose Self-Monitoring?

The aim of self-monitoring is to accumulate exact facts about blood glucose levels through the years at multiple points. It helps to maintain regular glucose levels and prevent hypoglycemia, and permits the subsequent to be scheduled accordingly:

- The treatment regime/insulin doses
- Dietary intake
- Physical activity.

Such glycemic control is vital in the prevention of the long-term complications of diabetes.

In addition to monitoring diabetes treatment effects and identifying blood sugar highs and lows, self-monitoring is a method that guides overall treatment goals. Self-monitoring also gives perception into how diet, exercise and different factors, such as illness and stress, have an effect on blood sugar levels.

Self-monitoring allows patients enhance their knowledge of glucose levels and the effects of different behaviors on their blood glucose.

Patients on glucose-lowering drugs can take their self-monitoring information to their health care provider, allowing them to measure prescriptions accordingly and advise any modifications to diet and exercising.

Strict glycemic control in type 1 diabetes is tough to achieve - in spite of correct training on self-monitoring, the most common measurement does not provide enough information to avoid hypoglycemia.

Who Should Self-Monitor Blood Glucose?

It was formerly only people with insulin-treated diabetes- type 1 mainly - who would be recommended to self-monitor their blood glucose levels. Global guidelines now state that there is enough evidence for the benefit of glycemic control to suggest self-monitoring to anyone with diabetes, together with those with type 2 diabetes who do not need insulin treatment, as long as there is sufficient healthcare guide. Adequate support includes the following:

- The monitoring is included into an education program to promote appropriate treatment according to blood glucose values
- There is shared management with health care providers to offer a clear set of instructions for acting on results.

The type of diabetes determines how often self-measurement is needed. Type 1 diabetes demands several day by day measurements while insulin-treated type 2 diabetes needs only around two a day. If no insulin treatment is needed, much less than daily measurement can be sufficient.

Target Blood Glucose Levels

The general goal of glycemic control for adults with diabetes has been set by the American Diabetes Association, whose guidance is followed by health care providers. It states:

- The HbA1c level (a marker of average glucose levels over recent months) should be reduced to 7% to lessen the risk of diabetes complications

- If possible, and as long as hypoglycemia can be avoided, some individuals may be able to target an HbA1c of 6.5%.

Less ambitious HbA1c goals (which include getting below 8%) are appropriate for some patients, inclusive of the ones who have any of the following:

- History of severe hypoglycemia
- Limited life expectancy
- Advanced diabetes complications
- Extensive coexisting conditions.

Much less stringent targets may also be suitable for people with long-standing diabetes who discover targets difficult despite disease control strategies.

The 7% HbA1c level informs the equivalent self-monitoring targets that patients can aim for (and again, much less ambitious targets are appropriate for some patients):

- Before meals (preprandial) - 70-130mg/dl (3.9-7.2mmol/l)

- After meal (postprandial, 1-2 hours after start of meal) - less than 180 mg/dl

How Is A Blood Glucose Monitor Used?

A glucose meter electronically reads a small sample of blood on a test strip. The blood is commonly drawn by means of a pores and skin prick at the tip of a finger.

Over 20 varieties of glucose meter are commercially available, varying in size, the quantity of blood needed and electronic memory and analysis features. While a few permit graphs to be computed, for many it is up to the user to maintain meticulous information along with details of times, diet and exercise. Practical hints for blood glucose monitoring include:

- Handle the meter and test strips with clean, dry hands
- Use the test strips precise for the meter and keep these in the original container
- Use a test strip only once and discard
- Strips may be calibrated with the meter for accuracy, and some meters require coding with each new canister of strips
- Take a look at for expiration dates
- Keep in a cool, dry place
- Take the meter to office visits for checks by providers.

Practical steps are also needed in preparation of the skin prick for a blood sample. The skin site must be cleaned with warm, soapy water and dried, or an alcohol pad may be used. In any other case - if foods have been handled recently, as an instance - false readings can occur.

The lancet sizes vary and can be adjusted to prick the pores and skin one and produce the different amounts of blood needed by using various meters. Thinner and sharper lancets are typically the most comfortable. Lancets need to no longer be reused after single use.

To reduce pains, the sides of the finger may be used and fingers can be rotated, such as any of the 5 digits in preference to the index finger or thumb.

At the same time as the most accurate measurements are enabled via using the fingertips or outer palm, some meters permit using different sites consisting of the upper arms and thighs.

When Should Glucose Self-Monitoring Tests Be Done?

Individual instances of diabetes require different levels of blood glucose monitoring. The frequency of testing can change for an individual as well; the frequency may also need to be intensified within the event of changes to medications, stress levels, diet or activity levels.

Examples of the type of facts that can be provided via meter readings include checking oral medicines or long-acting insulins through the use of night-time fasting blood glucose (FBG) readings, taken at around 3 or 4am. Test results from before ingesting can assist to guide changes to meals or medicines, and those obtained 1-2 hours following a meal are informative while learning how blood sugar levels are affected by foods. Tests at bedtime also help inform modifications to weight loss plan or medications.

Real-Time Continuous Glucose Monitoring

Non-stop glucose monitoring overcomes the hassle of taking several guide daytime readings from pores and skin pricks.

Patients with Type 1 diabetes usually do between four and eight finger-prick measurements every day, and rarely monitor night-time blood glucose levels.

Such self-monitoring can result in speedy modifications in blood glucose called excursion, together with postprandial hyperglycemia, asymptomatic hypoglycemia and fluctuations overnight.

Actual-time continuous glucose monitoring has been shown to be more powerful than self-blood glucose measurement in reducing hba1c in type 1 diabetes as it provide detailed information on glucose patterns and tends.

The most important crucial factor to the fulfillment of the devices is motivation and compliance of the consumer.

The available continuous monitors - some of which can be combined with insulin pumps-consist of an electrochemical sensor positioned below the skin and replaced every 3-7 days.

CHAPTER 15

MANAGING DIABETES WITH DIET & FOOD PLANNING.

Alongside exercise, a wholesome diet program is a crucial detail of the life-style management of diabetes, in addition to being preventive against the onset of type 2 diabetes.

Maintaining a very good food regimen is likewise a crucial a part of maintaining tight control of blood sugar levels, itself important for minimizing the threat of diabetes complications.

The good news to people living with diabetes is that the

circumstance does not preclude any specific type of food or require an uncommon food plan - the goal is much the same as it would be for everyone wishing to eat a healthy, balanced diet.

What DIET IS BEST FOR DIABETES?

Having diabetes does not demand any particular difficult nutritional demands, and at the same time as sugary meals obviously have an effect on blood glucose levels, the weight loss plan does not have to be completely sugar-free.

Nutritional issues vary barely for peoples with different types of diabetes. For peoples with type 1 diabetes, weight loss plan (diet) is about managing fluctuations in blood glucose levels while for peoples with Type 2 diabetes, it is all about losing weight and restricting calorie consumption.

For people with Type 1 diabetes, the timing of food is particularly essential in terms of glycemic control and in relation to effect of insulin injection.

In general, however, a healthy, balanced diet is all that is needed, and the benefits are not limited to good diabetes control - they also imply good heart health. A healthy diet normally includes a variety of fruits and vegetable, whole grains, low-fats dairy products, skinless chicken and fish, nuts and legumes and non-tropical vegetable oils.

The following are some general dietary suggestions for a healthy life-style:

- Eat frequently - keep away from the effects on glucose levels of skipping food or having behind schedule meal due to work or long journeys (take healthy snacks with you)

- Eat vegetables and fruits and devour them in place of high-calorie meals - a variety of fresh, frozen and canned is good, but keep away from high-calorie sauces and meals containing added salt or sugar

- Whole grains high in fiber are recommended as a healthy source of carbohydrate

- Consume pulses, a low-fat starchy source of protein and fiber, together with beans, lentils, chickpeas and lawn peas

- Reduce intake of saturated and trans fats by having poultry and fish without the skin and cooked, as an instance, under the grill, rather than fried

- Take a similar approach to cooking red meat while decreasing intake and looking for the leanest cuts

- Consume fish twice a week or more, however keep away from batters and frying - move for oily fish which include salmon, mackerel, sardine, trout and herring, which are rich sources of omega-3

- Keep away from partially hydrogenated vegetable

oils and limit saturated fats and trans fats - replace them with monounsaturated and polyunsaturated fats

- Dairy awareness helps reduce fats intake - select skim (fats-free) milk and low-fats (1%) dairy products, reduce intake of cheese and butter and swap out creamy sauces for tomato-based ones

- Cut back on sugar by avoiding added sugars in drinks and ingredients - have tea and coffee without sugar, keep away from fruit that is canned in syrup and take note of food labels

- Cut back on salt - put together foods at home with little or no salt and keep away from meals with high sodium such as processed foods

- Cut back on portion sizes - be cautious of quantities consumed when eating out

- Be cautious of "diabetic" foods - they are of no specific gain and can be expensive

- Drink alcohol only in moderation - as a guide, no more than one drink a day for women and no more than two for men.

Professional Help with Lifestyle Changes For Diabetes

In the US, the Community Preventive Services Task Force run diabetes prevention programs that help with enhancing diet for people at risk of, or newly diagnosed with type 2 diabetes. The program may consist of:

- Goals towards weight loss

- Individual and group education session on food regimen(diet) and exercise

- Meetings with diet and exercise counselors

- Individually designed diet and exercise plans.

Individuals in the national diabetes prevention program have access to a life-style coach to learn more about healthy eating and exercise.

Obesity, Diabetes and Diet

Obesity is a risk factor for Type 2 diabetes, and obesity in people who already have diabetes results in bad control of blood sugar, blood pressure and cholesterol levels.

Some other situation with being overweight or having obesity is that it can worsen many of the complications of the diabetes.

Weight loss can be accomplished by following the tips above and limiting the consumption of calories.

CHAPTER 16

MANAGING DIABETES WITH PHYSICAL ACTIVITY AND EXERCISE

According to the Centers for disease control and prevention (CDC), over 29 million people in the United State have diabetes - a condition in which the body doesn't make enough insulin (type 1 diabetes), or is unable to use insulin properly (type 2 diabetes).

Insulin is a hormone, made in the pancreas, which regulates blood sugar (glucose) levels, and allows the body to use glucose for strength.

Exercise can help reduce complications of diabetes which include:

- Heart sickness and stroke
- Blindness and other eye problems
- Kidney disease
- Amputations due to harm to blood vessels and nerves, leading to infection

A further 86 million people have pre-diabetes - a health condition that increases their risk of developing type 2 diabetes and different ailments.

Exercise and Diabetes

Aerobic sporting activities (exercises) such as brisk

walking and hiking can also help to manage the onset of diabetes symptoms.

Preventing the onset of diabetes for people with pre-diabetes or managing symptoms for those who have already got the condition, is crucial to keep health and prevent complications. Exercising is one proven way to help manage diabetes.

According to a joint position statement by the American College of Sports Medicine and The American Diabetes Association, exercise:

- Performs a key role in preventing and controlling blood sugar levels

- Can prevent or delay type 2 diabetes

- Can prevent diabetes in the course of being pregnant (gestational diabetes)

Staying physically active also helps prevent diabetes-associated health complications and improves overall quality of lifestyles.

Exercise is beneficial for people with diabetes as it improves insulin sensitivity by helping the cells of the body use insulin that is available. Physical activity additionally stimulates a separate mechanism, unrelated to insulin, to permit the cells to use glucose for energy, thereby regulating blood glucose levels.

Types of Exercise For People With Diabetes

The American Diabetes Association recommends two types of physical activity for those with diabetes: *Aerobic exercise and Strength training.*

Aerobic exercising

Also referred to as *cardiovascular exercise*, aerobic activity helps the body use insulin more efficaciously. It brings other benefits too, including:

- Strain remedy

- Progressed circulate

- Decreased danger of coronary heart disorder

- Lower blood strain

- Improved cholesterol levels
- Strong bones
- Weight management
- Better temper

Examples of cardio physical activities (aerobic exercise) consist of:

- Brisk strolling or trekking
- Low-effect aerobic exercise instructions
- Swimming
- Rowing
- Biking
- Basketball

- Dancing

- Skating

- Tennis

- Jogging

- Tai chi

How much aerobic activity is needed?

The president's council on health, sports and nutrition recommends:

Half-hour (30 minutes) daily of moderate physical aerobic activity as a minimum of 5 times weekly

This recommendation is for adults between the age 18-64. Adults with diabetes must have this in mind and work towards meeting this target.

People with a hectic schedule can also find it useful to do numerous shorter exercises totaling 30 minutes daily - studies suggests that the benefits received are similar to the ones related to one longer exercise.

Strength training

Blood sugar levels can be reduced by strength training, such as using free weights.

Strength training, or resistance training, helps in lowering blood sugar levels and increase insulin sensitivity. In addition, it increases resting metabolism and builds stronger bones and muscle tissues, decreasing the risk of osteoporosis.

Examples of strength training consist of:

- Lifting free weights
- Lifting heavy objects, such as bottles of water or canned food
- Weight machines
- Resistance bands
- Exercise that use body weight such as sits-ups, squats, planks, and push-ups
- Strength training classes

How much strength training is needed?

Strength training ought to be undertaken at the least twice every week, furthermore to the recommended amount of aerobic activity.

Stretching sports

Stretching exercises are crucial for everyone, together with people with diabetes. Stretching:

- Reduces the chance of injury from aerobic exercise or strength training

- Increases flexibility

- Prevents muscle pain

- Lowers stress levels

Incidental physical activity

It can be useful to consider incidental bodily activity-

everyday activity that are not classed as exercise however involve movement. A few studies shows that such activities can make a contribution to improved fitness. Types of incidental physical activities consist of:

- Taking the stairs instead of the elevator
- On foot (walking) to the bus stop
- Vacuuming
- Moderate intensity gardening
- Taking walks around the shopping center
- Washing the automobile

Monitoring Blood Glucose Levels When Exercising

To exercise adequately, many people with diabetes- especially those with type 1 diabetes or those on diabetes medicinal drugs - may need to check their blood glucose levels earlier than, at some point of, and after exercise.

This indicates how well the body is responding to exercise, and may help avoid blood sugar fluctuations, which can be dangerous.

Trying Out Blood Glucose Before, At Some Stage In, And After Exercising

Blood sugar levels must be tested half-hour before

exercising. If they are:

- Lower than 100 milligrams per deciliter (mg/dl) - blood sugar may be too low to exercise. Low blood sugar is called hypoglycemia.

- Between 100 and 250 mg/dl - this is the most excellent range, within which it is far safe for the majority to begin exercising.

- 250 mg/dl or higher - blood sugar may be too high to exercise. Perform a urine test for ketones (which indicate more insulin is needed to control blood sugar). This is typically only a concern for those with type 1 diabetes.

In the course of exercise, in particular long exercises or

new activities, blood sugar levels need to be examined every 30 minutes. Stop exercising if any of the subsequent symptoms are there:

- Blood sugar falls below 70 mg/dl
- Weakness
- Tingling
- Confusion

After exercise, check blood sugar level right away. Recheck levels numerous times over the following day – physical activity can lower blood glucose for as much as 24 hours.

Hypoglycemia and Exercise

If hypoglycemia (low blood sugar) is experienced during or after a exercising, it should be treated with straight away. This should be taking at least 15-20 grams of fast-acting carbohydrate such as:

- A sports drink

- Regular soda

- Glucose gel

- Jelly beans

Blood glucose levels should to be tested after 20 minutes, and the treatment repeated if they have not returned to normal. Follow the fast-acting carbohydrates with a protein such as peanut butter and crackers. Do not resume exercise until blood glucose returns to above 100mg/dl.

If hypoglycemia happens frequently throughout exercise, it can be important to adjust medications or the exercise regimen, or to absolutely consume a small snack before working out. Skipping food, strenuous exercise, or prolonged workouts can all cause hypoglycemia.

It should be noted that people with kind 1 diabetes are much more likely to experience hypoglycemia during or after exercise, even though people with type 2 diabetes may have problems if they are on medication for their condition.

When to See a Doctor

For people with diabetes, it is far recommended to consult with a healthcare expert before any exercising program commences.

It is miles advisable to consult a health practitioner before beginning any new workout program.

A doctor can advise on the impact of medications on blood sugar levels all through the period of activities, and can provide a target range for blood glucose levels in the course of exercises. They will give the best advice to the perfect time to exercise, primarily based on the affected patients individual schedule, meal plan, and medication.

A doctor may additionally perform a physical check-up,

looking at:

Heart health

Blood pressure

Diabetes-related complications

Depending on these complications, it may be recommended to keep away from strenuous activities, or particular sports.

It is also important to consult a medical doctor if hypoglycemia is experienced often at some stage in or after exercise, or if any other undesirable side effects are experienced.

Other Considerations

Beginning an exercise plan can be daunting. It is far crucial to:

- **Set realistic goals** - start slowly - with just 5-10 minutes of exercise every day - and gradually increase the frequency and intensity of the activity.

- **Include aerobic and strength based activity** - an exercise plan for diabetes management ought to include both aerobic exercise and strength training - studies shows that undertaking both forms of physical activity is more effective than doing just one of the two.

- **Take precautions** - constantly keep fast-acting

carbohydrates on-hand in case of hypoglycemia. Consider wearing a clinical alert bracelet in case of emergency.

- **Pick foot-wears wisely** - many peoples with diabetes have issues with their feet, because of poor circulation and nerve damage. Wear comfortable and supportive jogging footwear.

- **Be consistent** - to reap the advantages of exercise for diabetes, it has to be undertaken regularly.

Recommendations

1) How and Where to Buy Viagra Online Safely, Legally and Cheap: The Secret Behind How To Buy Viagra Online Safely Without A Prescription (With List Of Best Place To Buy Viagra Online) http://getbook.at/viagraonline

2) Viagra & Sildenafil: Uses, Dosage, Side Effects and Risks Information: The Secret Guide Behind How To Buy Viagra Online Safely, Cheap and Legally (With Best Online Pharmacy for Generic Viagra) http://getBook.at/viagra

3) Erectile Dysfunction (ED): Symptoms & Causes, Diagnosis, Treatment Online, And More Using

Viagra Without A Prescription (Including Where To Buy Viagra, Cialis, Levitra etc Drugs Cheap & Safely Online

http://getBook.at/erectile

4) Innovative Visualisation: The Power of Mind Perception -- GET MORE DONE THROUGH MIND MANIPULATION, INCENTIVES, PSYCH TRICKS AND MORE

http://getBook.at/innovative

5) Natural Healing and Remedies Cyclopedia: Complete solution with herbal medicine, Essential oils natural remedies and natural cure to various illness. (The answer to prayer for healing)

http://getBook.at/naturalhealing

6) **100 BEST CAT WELLNESS FOOD, DIET & RECIPES**: The hidden healing power diet for cat kidney problems, cat weight-loss, & pregnant cat diet; including recipes for all cat diseases and illness http://myBook.to/catfood

7) **The Brain, Mind and Memory Therapy:** The Science of embracing Change, Boosting Brain Power, Increasing Your Energy and Mental Strength. http://getBook.at/brainbook

8) **What Wikipedia Can't Tell You About Achieving Your Goals**: Why your objective setting never works out the way you plan http://getBook.at/wikipedia

9) **The First Year From Childbirth and beyond:** Inside-out Information on what to expect the first year and beyond early childhood for mothers and fathers made simple

http://getBook.at/childbirth

We love Testimonies, and we want to know how thus our publications have been of immense help to you. And please consider writing to us at www.engolee.com

Follow us on Social media at:

Website: www.engolee.com

Facebook Page: www.facebook.com/engolee

Twitter Page: www.twitter.com/engolee

About the Author

Dr. Jane A. McCall is a willing Health Researcher who is committed to blessing human race. She has developed a series of fabulous and highly effective healthful strategies and exercise programs. She applies her encyclopaedic knowledge and astonishing perception to analyze the background and underlying causes of various diseases affecting people in the world and then designs individualized and totally effective strategies to attain the desired results in solving human related problem with diseases. Jane is totally committed to helping the world discover their ideal expression of complete wellbeing.

Acknowledgments

The Glory of this book success goes to God Almighty and my beautiful Family, Fans, Readers & well-wishers, Customers and Friends for their endless support and encouragements.

www.ingramcontent.com/pod-product-compliance
Lightning Source LLC
Chambersburg PA
CBHW021402210526
45463CB00001B/194